WE HAVE A DREAM

written by
'BIRDGIRL' DR MYA-ROSE CRAIG

illustrated by
SABRENA KHADIJA

MAGIC CAT 🐱 PUBLISHING

'BIRDGIRL'
DR MYA-ROSE CRAIG

is a British—Bangladeshi naturalist, environmentalist and race activist. In February 2020, Mya-Rose became the youngest British person to be awarded an honorary doctorate in science from Bristol University, in recognition of her five years of campaigning for diversity in the environmental sector.

SABRENA
KHADIJA

is a Sierra Leonean—American illustrator. As a Black creative, Sabrena takes a lot of pride in creating work that helps others feel seen and inspired. Not only to see beauty within themselves, but to recognize and acknowledge that of others. Of her practice, she says, "I am one of the growing number of human beings who seek inclusive and innovative spaces to explore art and creativity in meaningful and impactful ways."

To all the young activists with big dreams and to my family for helping me achieve mine. — M - R . C .

To my little brother Ahmad. I can't wait to watch your brilliance illuminate the world. — S . K .

MAGIC CAT PUBLISHING

We Have a Dream © 2021 Magic Cat Publishing Ltd
Text © 2021 Dr Mya-Rose Craig
Illustrations © 2021 Sabrena Khadija
Photograph of Dr Mya-Rose Craig by Oliver Edwards Photography
First Published in 2021 by Magic Cat Publishing Ltd
The Milking Parlour, Old Dungate Farm, Plaistow Road, Dunsfold, Surrey GU8 4PJ, UK

The right of Dr Mya-Rose Craig to be identified as the author of this work and Sabrena Khadija to be identified as the illustrator of this work has been asserted by them in accordance with the Copyright, Designs and Patents Act, 1988 (UK).

A catalogue record for this book is available from the British Library.

ISBN 978-1-913520-20-5

The illustrations were created digitally
Set in Bakerie, Lato and Supa Mega Fantastic

Published by Rachel Williams and Jenny Broom
Designed by Nicola Price
Edited by Helen Brown

Manufactured in Lithuania, BAL0521

9 8 7 6 5 4 3 2 1

MIX
Paper from responsible sources
FSC® C107574
FSC
www.fsc.org

MY DREAM

INDIGENOUS PEOPLE AND PEOPLE OF COLOUR ARE DISPROPORTIONATELY AFFECTED BY CLIMATE CHANGE. AND YET THEY ARE UNDER-REPRESENTED WITHIN THE ENVIRONMENTAL MOVEMENT.

NOT ANYMORE.

I BELIEVE THAT TO PROTECT THE ENVIRONMENT IS TO LEVERAGE THE INPUT AND CONTRIBUTION OF AS MANY PEOPLE AS POSSIBLE. BUT IT IS NOT FOR ME TO SPEAK FOR OTHERS.

THE TIME HAS COME FOR PEOPLE TO SPEAK FOR THEMSELVES.

THE COVID-19 PANDEMIC HAS PROVED THAT GOVERNMENTS ARE WILLING TO ACT IN EXTRAORDINARY WAYS TO PROTECT THE WELL-BEING OF THEIR CITIZENS. THE IMPENDING ENVIRONMENTAL CRISIS DEMANDS NO LESS ACTION. DURING LOCKDOWN, I SPOKE TO THIRTY YOUNG CAMPAIGNERS FROM INDIGENOUS COMMUNITIES AND COMMUNITIES OF COLOUR EXPERIENCING THE STARK REALITY OF OUR CHANGING PLANET. OUR CONVERSATIONS MADE CLEAR A UNITED DREAM.

A dream for climate justice.

A dream for a healthy planet.

A dream for a fairer world, for all.

'BIRDGIRL' DR MYA-ROSE CRAIG

BRITISH–BANGLADESHI NATURALIST, ENVIRONMENTALIST AND RACE ACTIVIST

AUTUMN PELTIER

BORN: 2004
ETHNICITY: MANITOULIN WIKWEMIKONG FIRST NATION

AUTUMN HAS ALWAYS UNDERSTOOD THE IMPORTANCE OF CLEAN WATER and the need to protect it. At just eight years old, she was attending water ceremonies on First Nations reserves. Now the Chief Water Commissioner for the Anishinabek Nation, a position previously held by her great-aunt and mentor Josephine Mandamin, Autumn is an impressive voice for the universal right of clean drinking water for First Nations communities and across Mother Earth.

Autumn spent her childhood on the shores of Lake Huron, one of the largest freshwater lakes on Earth, and comes from the territory of Wikwemikong, a First Nations reserve on Manitoulin Island in Canada. Autumn realized the importance of clean water after taking part in a ceremony in a village that had no access to it. "People my age didn't even know what a drinking water tap was. They had to boil water in big pots," Autumn explains.

In the Global North (the richest and most industrialized countries), getting clean water is something that most people may not pay attention to, but this is not the case everywhere in the world. Canada is one of the world's wealthiest countries, but there are First Nations communities who aren't able to drink their water as it's contaminated by pollution and oil pipelines. Autumn thought this was unacceptable.

Campaigning has been in Autumn's family for a long time, and her great-aunt Josephine Mandamin was known as the 'water walker'. Josephine was the Chief Water Commissioner for the Anishinabek Nation until she died in 2019. Autumn reveals, "Even though my great-aunt is not with us anymore, she's still my biggest mentor." Soon after, Autumn was appointed the Chief Water Commissioner and became known as the 'water warrior'. Why? Because when addressing the United Nations General Assembly, Autumn told them to "Warrior up!"

Autumn lends her voice to the most vulnerable communities when speaking to people in positions of power. In 2016, she met Canadian Prime Minister Justin Trudeau and expressed her upset about his decision to allow an oil pipeline project to proceed. The Assembly of First Nations Youth Council was so inspired by Autumn's bravery that it created a fund to help protect water for future generations.

Autumn thinks it's important for people to speak up for the next generations, "as it can make a huge impact," she says. This is a global issue, and Autumn encourages everyone to join the fight.

> "CLEAN DRINKING WATER is a basic HUMAN RIGHT. Nobody should have to go without it."

VANESSA NAKATE

BORN: 1996
ETHNICITY: MUGANDA

GROWING UP IN KAMPALA, UGANDA,

Vanessa has seen first-hand the impact of climate change. After founding a movement for young people in Africa, Vanessa was catapulted into social activism after being cropped out of a photograph with four white climate activists. Instead of being pushed to the side, she spoke out against discrimination. Vanessa's courage is inspiring and we share a dream of all climate activists having their voices heard.

After graduating from university, Vanessa was surprised to learn that climate change was contributing to the poverty in her community in Uganda. Climate change and rising temperatures were causing periods of low rainfall which led to limited access to food and clean water.

So, Vanessa organized a strike against the government's inaction on climate change. For several months she was the only protester standing outside shops, petrol stations and even the gates of Parliament. But not for long. Soon people responded to her calls on social media, and Vanessa founded Youth for Future Africa.

In January 2020, Vanessa attended a conference with other climate activists, including Greta Thunberg. When a photograph of the event was published, Vanessa had been cropped out and the picture only showed four white climate activists. In response, Vanessa tweeted: "You didn't just erase a photo. You erased a continent. But I am stronger than ever." She received messages of support from other climate activists who had similar experiences, but had lacked the courage to speak out. Ever since, Vanessa has felt a greater responsibility to amplify their voices because if they are not heard, climate justice cannot be achieved.

Vanessa is passionate about uplifting marginalized climate activists because, she says, "It is their communities that are suffering the most from climate change." Countries in the Global North, such as the United States, are the biggest contributors to climate change, but they experience the effects of it the least. For the Global South, including countries like Uganda, it is a much greater problem for a lot of the people who live there.

Many climate activists want to change the lives of the people in their communities, but this is not possible if they are not given the platform or resources to do that. Vanessa embraces everyone's story, and uses shared experiences as a source of energy to keep pushing for climate action.

> "CLIMATE JUSTICE has to include EVERYONE. Every person, every country."

LESEIN MUTUNKEI

BORN: 2004
ETHNICITY: KENYAN MAASAI

WHEN FOOTBALLER LESEIN learned about the devastating impact of trees being cut down in his home of Kenya, he combined his love of football with his desire to protect his country's forests. His initiative, Trees 4 Goals, was simple: for every goal he scored, he would plant eleven trees, one for each team member. Lesein's practical activism is admirable; he has used his passion to inspire others to play a key role in fighting climate change.

Did you know that in the time it takes to say 'deforestation', a section of forest the size of a football pitch is destroyed? Lesein was twelve years old when he found out that Kenya was losing 500 square kilometres of forest each year, the equivalent of 164 football pitches every day. Lesein loved to play football in the forests near his home, going almost every day, and realized they could disappear if nothing was done.

Planting trees is a great solution to the climate crisis as they absorb carbon dioxide in the air as they grow. Lesein started small and began planting trees at his grandmother's house in Nairobi, the capital of Kenya. But soon he wanted to make a bigger impact, which meant a bigger

commitment. He loved football and so for every goal he scored, he decided to plant a tree, calling his initiative Trees 4 Goals.

Then, Lesein thought, as every goal is a collaborative team effort, what if for every goal scored eleven trees were planted, one to represent each player on the team? His football team loved the idea, and his school adopted it in all of their sports teams. Soon, together they had planted almost 1,000 trees!

Lesein's school was so impressed with his achievement that they nominated him to attend the first United Nations Youth Climate Summit. At fifteen years old, Lesein was one of the youngest people there, and when he returned, he planted a tree with the President of Kenya.

This experience gave Lesein the confidence to think even bigger. He is now passionate about the possibility of an eco-friendly society. "It is up to us to do the best we can to give the planet to our children so they can experience the beautiful nature we have," he says. He plans to reach out to famous football players to help create a forest of Trees 4 Goals in each county in Kenya, then a forest of Trees 4 Goals in each country in Africa. Lesein believes that football can connect, engage and inspire young people to take action to achieve a greener future.

"TREES help us in so many ways — even cleaning the AIR we breathe. The more we can PLANT, the BETTER!"

SUMAK HELENA SIRÉN GUALINGA

BORN: 2002
ETHNICITY: ECUADORIAN INDIGENOUS, AMAZONIAN KICHWA SARAYAKU, FINNISH

HELENA HAS STOOD AGAINST THE INTERESTS OF BIG OIL COMPANIES and their environmental impact on Indigenous land for profit since childhood. She grew up as part of a small Indigenous community in the Ecuadorian Amazon – one that faces the constant threat of losing its land – which brought her into the world of being an Indigenous rights and rainforest defender. Helena uses her notable platform to speak out and bring awareness to these important issues.

Helena comes from the Kichwa Sarayaku community, which is located in the Ecuadorian Amazon. In 2002, the year Helena was born, a company entered Sarayaku looking for oil. As a child, Helena witnessed the impact this had on her community. "It has always been a part of my life that people were fighting for our communities. . . that someone was trying to take our home from us," she reveals.

Helena was raised in a family of strong women. Her mother was the president of the Kichwa Women's Association, and her sister, aunt and grandmother are defenders of Indigenous women's rights. It therefore felt natural for Helena to join them. The Sarayaku community fought hard and were able to protect their forest because, Helena says, "We never gave our consent for them to come to our community and the government didn't let us have a say." Helena realized the government made the decisions on their behalf; their voices were not heard.

As her father worked at a university in Finland, Helena spent her childhood living between two homes, and has been able to broadcast her community's message on a wider scale. She uses her voice and platform to raise awareness about what is happening to Indigenous people and how it is connected to the climate crisis.

In 2020, Helena started Polluters Out, a youth movement attempting to remove the influence of the fossil-fuel industry within Indigenous lands. "The most important step towards stopping climate change is to stop extracting fossil fuels," she explains.

Helena encourages people to listen to marginalized communities. The more that speak up and are heard, the more likely it is we can create a healthier, more peaceful world for everyone.

> "My community has to keep FIGHTING to protect our FOREST, our ANIMALS and our PEOPLE from the threats to everything we know as HOME."

GHISLAIN IRAKOZE

DREAM:
A WASTE-FREE WORLD

BORN: 2000
ETHNICITY: BLACK RWANDAN

AS A CHILD IN KIGALI, RWANDA,
Ghislain was witness to the densest landfills in the city. Landfills can be huge environmental hazards as many toxic substances can leach into the ground over time. With a dream of creating a waste-free world, Ghislain built a mobile app, Wastezon, to connect people to recycling points. His work is inspiring as it involves humanity and technology coming together to protect the planet.

Ghislain discovered the weight of the waste problem at eleven years old when he was volunteering on a school project at a local landfill site. Ghislain and his friends were researching the impact the toxins were having on nearby homes, when suddenly some rubbish came loose and went sliding down. Ghislain's friend was sadly hit and spent two weeks in the hospital recovering. The accident taught Ghislain something valuable. "In a waste-free world, this would have never happened," he says.

After this traumatic event, Ghislain started an environmental awareness project which empowered 200 students with hands-on skills of recycling plastic waste. Together, they recycled more than 100 tons of plastic, which resulted in 2,000 tons of carbon dioxide emissions being diverted from landfills.

But Ghislain didn't stop his fight there. He was more convinced than ever that it was possible to have a world without plastic waste – if only he could reach more people. His vision was to combine environmental activism and technology to prevent recyclables being dumped in the first place and help people to rethink their consumption habits, whether it was recycling, reusing or repurposing.

So, in 2018, Ghislain launched Wastezon – a mobile app connecting people to recycling points, giving incentives for them to recycle rather than dumping or burning their waste. There are now hundreds of households using the app and Ghislain has prevented a further 3,000 tons of carbon dioxide emissions being released into the atmosphere. And, as households upload their rubbish to be recycled, they become ambassadors encouraging others to join in too.

Wastezon's next step is to introduce a 'Smart Bin' that responds to both households' waste collection needs and the growing demand for compost. Thinking beyond his hometown of Kigali, Ghislain plans to bring out innovations that help communities throughout Africa dispose of their waste responsibly. As he acknowledges, "It's a global fight – a fight for our shared planet."

"I believe that a world without WASTE is possible. There should be no LANDFILL at all."

BRIANNA FRUEAN

BORN: 1998
ETHNICITY: SAMOAN

GROWING UP ON THE ISLAND OF SAMOA

in the South Pacific inspired Brianna to protect her home from rising sea levels. When sea levels rise, some people have to migrate to higher ground or become vulnerable to flood risk. We share a dream to fight climate change to help stem tides rising in Bangladesh where my family live, in Samoa where Brianna's family live, and for every family around the world.

Brianna's climate activism journey started when she was eleven years old. She attended a climate change workshop and her interest was immediately sparked, so she decided to set up an environmental group. At such a young age, Brianna was motivated by love for her island, which she felt was filled with richness.

Protecting her island and helping her community has been the inspiration behind Brianna's work. Climate change is the greatest threat to the Pacific, and the island of Samoa, along with other vulnerable communities around the world, is experiencing the effects of the planet heating up. As temperatures increase, ice begins to melt and runs into the oceans, causing the sea level to rise. Rising waters mean that islands end up flooded, or sometimes underwater. And it's going to get worse for future generations if we don't act. But Brianna is passionate about not looking at the climate crisis as the end of something, but instead viewing at it as the beginning of something better.

Brianna believes that young people today, by trying to build a bigger climate movement, will be the generation repairing the damage. "We can fix this as the upcoming generation of leaders with the ability, time, passion and energy," Brianna explains. She continues to strengthen the voice of young people in the Pacific through grassroots projects. She visits schools to raise awareness of climate change and empowers children to take action, both at home and globally.

Brianna describes herself as campaigning for "climate justice", rather than climate action alone. Climate justice addresses the climate crisis while also protecting human rights, such as access to housing and food.

Now studying at university in New Zealand, Brianna continues to fight for a climate conversation where all people feel welcome and valued. The climate space is a place for everyone and, when the weight of the world feels heavy, Brianna understands we need a lot more hands to carry it.

> "We must build a GLOBAL FAMILY. We must value EVERYONE the same, value everyone's HOME the same."

TALISSA M. SOTO KRENTZIEN

BORN: 1993
ETHNICITY: LATINX

TALISSA IS A GRASSROOTS ENVIRONMENTAL ACTIVIST, trainer and poet working for climate justice in the Netherlands. She organizes events where people can share their grief and anger about environmental loss, along with their visions for building a sustainable, just world. Talissa's fight for climate justice is a cause close to my heart as I believe both people and the planet should be treated fairly.

Born in Venezuela, Talissa became involved in environmental activism at the age of nineteen while studying at university in the United States. She had already been an activist focusing on racial and migrant justice, but became aware of the links between Indigenous peoples' rights, racism and environmentalism.

So, she began organizing gatherings, attending protests and offering training to climate activists. She also co-founded Climate Liberation Bloc (CLuB) to create a space for those most impacted by the climate crisis: Indigenous people and people of colour. CLuB organized workshops and talks which were safe spaces for activists to share their story.

Talissa fights for climate justice, a form of environmental action which also protects human rights. Talissa reflects on when a fellow activist asked her, "What good would it be to have a liveable planet if, when we walked down the street, we are subjected to a racist attack?" She realized that we cannot just create a liveable and sustainable planet, but it needs to be one in which everyone has the freedom to enjoy that planet, too.

To face the climate crisis, we have to challenge ourselves, but also, as Talissa says, "People who have been marginalized have the power to be fearless and loud," adding, "Do not allow the fear of being silenced stop you from saying what's important to say."

To promote open dialogue, Talissa liaises closely with other groups that are doing feminist and anti-racist work to bring it into the climate movement. The hard work has been worthwhile as Talissa sees more Indigenous people and people of colour taking centre stage in the European environmental movement. We are finally focusing on social issues and acknowledging that marginalized people need to be at the centre, which gives the next generations courage. As Talissa says, "Take the space that you need, because the planet needs you."

> "We need to have a movement that is willing to let everyone SPEAK, to create SPACE, to LISTEN to different voices, to be CHALLENGED."

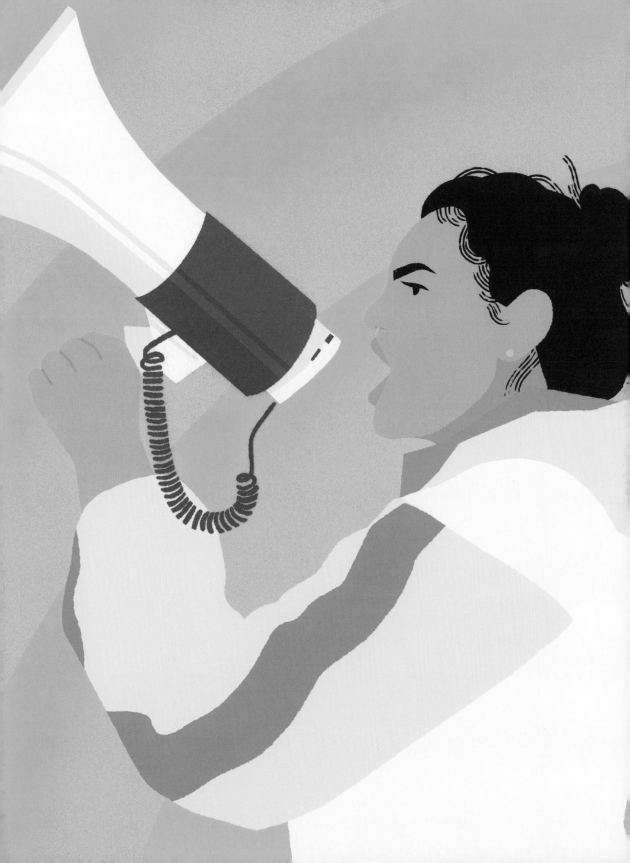

THOMAS TONATIUH LOPEZ, JR

DREAM: INDIGENOUS RIGHTS AND EARTH JUSTICE

BORN: 1992
ETHNICITY: OTOMI, DINÉ, APACHE, SICANGU LAKOTA

THOMAS SPENT MANY MONTHS PROTESTING at the Sacred Stone Camp on the Standing Rock Reservation in North Dakota to stop the construction of a dangerous oil pipeline. He then co-founded the International Indigenous Youth Council (IIYC) to fulfil a vital dream of Indigenous rights and environmental justice for us all.

Thomas was born and raised in Denver, Colorado, in the United States. He is the grandson of Chief Leonard Emmanuel Crow Dog, Sr and the son of Water Woman Sharon Dominguez and Sundance Chief Thomas Lopez, Sr. After graduating from university, Thomas heard about an Indigenous youth group which was organizing a cross-country 'spiritual run' to protest the construction of the Dakota Access Pipeline.

The Dakota Access Pipeline was to extend 1,900 kilometres, crossing through communities, tribal land and wildlife habitat to carry 570,000 barrels of oil. The protesters argued the pipeline would contaminate drinking water and damage sacred burial sites. As the son of a Chief, Thomas felt a duty to join the protesters at the Sacred Stone Camp.

For years, demonstrations took place along the road leading to the pipeline's construction site. Protesters waved flags representing over eighty different tribal nations. There were strict rules that the action would be peaceful. Even as the protesters found themselves under physical threat from the police, they remained calm, with Thomas saying, "Though you have brutalized us, we will not brutalize you." Decisions are still being made about whether the pipeline should proceed.

During Thomas's time at the Sacred Stone Camp, he lived in a traditional Indigenous community. For the first time, he was with people who had similar experiences to him. Thomas wanted to inspire people who identify as two-spirit, or Indigenous people who embody both a masculine and feminine spirit within them (though the term varies between nations), to reclaim their voices. So, he co-founded the IIYC to create a safe space to do so.

Thomas is adamant he won't stop his fight for Indigenous rights and environmental justice, as he is aware of the struggles of his ancestors. "If we want to be seen, we must make our presence known. Be the voice of the future and speak for the voices of the past," he says.

"We are WARRIORS. We are WATER PROTECTORS. We are LAND DEFENDERS. We demand JUSTICE for the EARTH."

REBECA SABNAM

BORN: 2003
ETHNICITY: NOAKHALI BENGALI

REBECA STARTED HER ACTIVISM JOURNEY AT THE AGE OF TEN after joining Cafeteria Culture, a non-profit organization working to achieve zero-waste schools. Alongside campaigning against waste, Rebeca advocates for Bangladeshi people and the issues they face as a result of climate change. As a Bangladeshi woman, I share Rebeca's dream for the climate crisis to be addressed as both an environmental issue and a human rights issue.

Rebeca spent her early childhood in Dhaka, the capital of Bangladesh. Bangladesh is one of the countries most vulnerable to the climate crisis and is often affected by flooding. As the country has less money to spend on defences, when floods occur, Bangladeshi people can lose their homes and livelihoods.

As a child, Rebeca remembers being carried to school on her uncle's back during severe floods. But other times, the school could be closed and wells were submerged, which meant no access to education and clean drinking water – two basic human rights. Rebeca migrated to the United States when she was six years old, and has made it her mission to highlight how climate change intersects with vulnerable countries.

In September 2019, Rebeca joined more than 200,000 people marching in Manhattan to demand climate action. She stood in front of a huge crowd and shared her experience. Her powerful speech was picked up by the media and, with flooding worsening, Rebeca used this new platform to call on people in positions of power to protect Bangladesh's poorest communities.

To help the climate crisis, Rebeca also campaigns for zero-waste schools. Working for Cafeteria Culture, the environmental education organization behind New York's decision to eliminate the 860,000 Styrofoam trays used every day in schools, Rebeca educates her classmates about the benefits of zero waste and has introduced plastic-free lunch days in her school.

But her hard work doesn't stop there – she is working to implement policies that mandate plastic-free lunch days in all New York schools. After successfully advocating with her classmates to get the New York plastic bag fee bill passed in 2020, there's no doubt that Rebeca will achieve anything that she sets her mind to.

> "Bangladesh exemplifies how interconnected the CLIMATE EMERGENCY is to RACIAL JUSTICE and POVERTY."

LITOKNE KABUA

BORN: 2002
ETHNICITY: MARSHALLESE

LITOKNE LIVES WITH THE IMPACTS OF GLOBAL WARMING (the rising temperature of Earth) in his home in the Marshall Islands. The marine landscape is declining at a dramatic rate and, as the ocean gets warmer and coral reefs are bleached, his community's homes and livelihoods are lost forever. To protect his island, Litokne has impressively joined a group of sixteen activists in petitioning the United Nations to engage with the destructive effects of climate change.

Litokne lives on Ebeye Island which is one of the five coral islands and twenty-nine atolls (a ring-shaped island formed of coral) that make up the Marshall Islands, where more than 10 per cent of the world's atolls can be found. Thousands of marine animals depend on coral reefs for survival, including the fish which Litokne's family eat.

Along with protecting their food source, Litokne campaigns to look after his community's safety, too. Coral reefs are natural barriers that absorb the force of waves to keep people safe from storms. But as the reefs are destroyed by climate change, waves are reaching islands and washing away homes. Almost 60 per cent of Litokne's community has been affected by this, and it's getting worse.

Coral reefs are also bleaching as the water gets hotter. It can only take 1 degree Celsius of warming to cause ecosystems to break down. Coral can also bleach for other reasons, such as pollution or too much sunlight. All of these triggers are caused by the climate changing – something which Litokne is campaigning to stop.

Litokne has joined a group of sixteen youth activists in petitioning the United Nations to combat climate change. Under the United Nations Convention on the Rights of the Child, countries have to protect children and because of the threat of the climate crisis, these conditions are not currently being met.

Litokne explains that if we take action now, "We can reduce climate change to a significant level, allowing us to live a sustainable life." Currently, the Marshall Islands are struggling and without immediate change, Litokne's home could soon be underwater. But in an effort to preserve his island, Litokne uses his platform to implore others: "My beautiful home is only two metres above sea level. We have only two metres of our culture left, only two metres of our future left. Will you help us?"

> "Coral reef is suffering from mass CORAL BLEACHING as a direct result of CLIMATE CHANGE."

TYRONE SCOTT

BORN: 1991
ETHNICITY: ITALIAN–GERMAN, BRITISH AFRICAN–JAMAICAN

TYRONE IS NO STRANGER TO TAKING ACTION, whether protesting with the school climate strikes, striving for social housing or writing about climate justice. He will not stop until world leaders prioritize the global climate emergency over profit. As a fellow campaigner, Tyrone passionately raises awareness of the climate crisis.

Born in Germany to an Italian–German mother and a British African–Jamaican father, Tyrone believes his multicultural background has contributed to his concern for a crisis which is global and affects us all – climate change. Now living in London, Tyrone became actively involved in climate activism in his late teenage years.

Tyrone joined the Green Party, a political party that campaigns for a greener, more sustainable way of living, and believes the planet should come before making money. Tyrone spent many days and nights demonstrating on the streets of London. "The main thing any one person can do is to join the movement and take action," he says. So, in 2019, Tyrone stood as a candidate to represent the area where he lived – Hackney South and Shoreditch – in the parliamentary election. Tyrone could no longer sit back and allow the effects of climate change to get worse. His goal was to bring his community a more environmentally friendly future.

Tyrone believes in the importance of tackling social and environmental issues, and works as a community organizer for a national housing charity. He campaigns for more social housing and demands that these homes should retain 95 per cent of energy – which is good for the planet and also reduces energy costs. He has seen first-hand how innovative social housing solutions can both lower carbon emissions and tackle inequality.

Alongside campaigning, Tyrone is also a journalist. He writes about the intersectionality between climate and racial justice, diversity in the environmental movement and his own experiences as a climate activist. By calling out the lack of diversity within environmental organizations, Tyrone puts pressure on these movements to be more inclusive.

Speaking about his peers, Tyrone is impressed at how engaged in politics and climate change young people are today. It gives him real hope, as he acknowledges, "The future is so progressive and we're going to change the world!"

"We must stop people EXPLOITING non-natural resources for PROFIT. The PLANET must come FIRST."

ARIEL CHEN

BORN: 2004
ETHNICITY: HAN CHINESE

ARIEL HAS BEEN A BIRDER since the age of twelve and, just like me, birdwatching means she views the world through a nature conservation lens. To protect the birds she loves, Ariel began campaigning – starting a blog, writing articles and setting up a birding club. These efforts were to encourage her community in China to notice the different wildlife living around them and to understand how essential it is to live together in peace.

Active birder Ariel has seen over 1,500 species of birds and calls herself a "birding fanatic". She loves China's vast and diverse landscape and how it's home to an abundance of wildlife. Ariel first became involved in environmental activism at the age of thirteen while attending school in Beijing, the capital of China.

Encouraged by her parents, she decided to share her passion for her country's wildlife by starting the Birding Club and invited classmates to join. Ariel taught lessons, hosted off-campus birding trips and organized lectures with experts – all of which influenced others to notice the natural world around them. The club members debated topics like why humans should protect animals and the importance of nature conservation.

Alongside the Birding Club, Ariel began an Animals Around project. Using cameras to record creatures in their natural habitat, Ariel presented the footage to classmates with the aim of further promoting awareness of the beauty of wildlife. She hoped that they would notice the different wildlife living around them, understand how they could coexist and how the relationship between them and animals can improve. "I'm aware that the world is not only for humans to live in but also for animals and we must think about them as we live our life and, most importantly, how our actions might affect them," she says. "We must use our efforts to protect them."

Ariel's next step is to talk to people who reside in remote places, like those living in relatively poor, rural communities. These locations have great biodiversity around them, and Ariel understands that different communities have different needs. Some may believe that earning money is more important for them than protecting the environment, so it's crucial to talk and listen, and then take the best course of action.

Fortunately, Ariel is not alone in her endeavours. She is part of a generation engaged in wildlife conservation across China, giving hope to a country with such rich fauna.

> "Wildlife should be able to COEXIST in the same space as humans in a SUSTAINABLE relationship."

ZANAGEE ARTIS

DREAM: GLOBAL CLIMATE JUSTICE

BORN: 1999
ETHNICITY: BIRACIAL

ZANAGEE CO-FOUNDED ZERO HOUR,
a youth-led climate justice organization, after becoming frustrated that young voices were being ignored in the conversation around climate change. From the United States, Zanagee's passion about the leadership of youth has roused a movement, and he has impressively organized global strikes, protests and rallies to provide a platform where people can come together and have their voices heard.

Zanagee has always felt connected to the ocean, living on the coast of Connecticut, and as a child dreamed of becoming a marine biologist. Inspired by his activist parents, he began researching sustainability and realized the impact of waste on the environment, especially in the sea.

When Zanagee was seventeen years old, he started a group at school to reduce waste. He wanted classmates to start recycling paper, to compost their food waste and to use less plastic. But, Zanagee admits, "It was hard to make change without engaging my peers."

So, he started a sustainability committee. The committee raised enough money to install water bottle refilling stations throughout the school, which meant people no longer bought plastic water bottles from the canteen. It was clear to Zanagee that change occurred because people came together, which he still believes is important in his work today.

And so when Zanagee met the other Zero Hour co-founders – Jamie Margolin, Nadia Nazar and Madelaine Tew – during a three-week summer programme at Princeton University, they came up with an idea to create a "youth-led climate justice organization to organize youth climate marches". Zanagee became the logistics director, in charge of submitting permits for marches, estimating attendance numbers and helping sister marches.

Zero Hour's main aim was to centre young people, most significantly Indigenous people and people of colour, who were working within the environmental movement, but whose voices were not being heard. Together, they built a movement around the intersectionalities between systems of oppression that continue to cause climate change, including racism, and strive to uplift voices environmentally impacted by these systems. Through his efforts, Zanagee has inspired young climate activists from around the world to speak out, speak loudly and speak proudly.

"We are RACING against the clock. This is our chance to get people to LISTEN. To CHANGE."

SULTAN AHMED

BORN: 1995
ETHNICITY: CHANDPURI BENGALI

SULTAN IS A WILDLIFE CONSERVATIONIST

working to protect species in Bangladesh. He plays an important role in the Global Youth Biodiversity Network (GYBN), inspiring young people to unite for the sustainable use and conservation of biodiversity for a healthy environment. My mother's family are Bangladeshi, so I feel connected to Sultan's journey. In his words, "We need to work together to preserve life on Earth."

During Sultan's childhood in rural Southern Bangladesh, his father taught him how vital trees are – how they provide oxygen for us and give homes to wildlife. When Sultan turned twelve, he created a nursery to grow seedlings and set planting goals so his trees could offer food and shelter for animals. It was important work as deforestation is one of the most significant drivers of biodiversity loss in Bangladesh, along with climate change.

Sultan's parents instilled in him a love of learning and the environment, and so he decided to study wildlife conservation at university. He discovered that a range of animals will face extinction in Bangladesh due to climate change, including many species of beautiful butterflies. Sadly, already a quarter of Bangladesh's wildlife is under threat while rapid destruction of their habitats continues. "Human beings have reached a point that has enabled us to inflict damage on our natural world – driving species extinctions," Sultan says.

Sultan has been a dedicated climate activist for years, but hasn't been campaigning alone. He works as Bangladesh's youth representative for Global Youth Biodiversity Network (GYBN). The GYBN is an international network of young people from across the world who come together to prevent the loss of biodiversity and to preserve Earth's natural resources. Sultan shares a vision to transform the world into one where people live in harmony with nature.

Campaigning to encourage other young people to conserve biodiversity and promote sustainable tree planting, Sultan passionately expresses how, "Biodiversity is life; it is crucial to the balance of ecosystems and the survival of all species."

Sultan is determined to play a part in a generation that is conscious of, and educated about, the environment, who will seek to educate others and, ultimately, help conserve such a biodiversity-rich planet.

"We must think GREEN: connect with NATURE, end species extinction and PROTECT all life on Earth."

NEJMA ALI MEHIDI

BORN: 2002
ETHNICITY: FRENCH, ALGERIAN ARAB

AT SEVENTEEN YEARS OLD, NEJMA REALIZED SHE WAS THE FIRST GENERATION

to know we're hurting the planet, but potentially the last generation that can do anything about it. So, she became a climate activist, engaging in global climate strikes to take back some agency on her future and on behalf of future generations. I share Nejma's dream of a fair world – one where people are equal and human beings globally stand together to combat climate change.

Born in France, Nejma was raised to look at and think about everything. Therefore, it wasn't a surprise to her parents that she decided to explore the environment through the medium of photography while studying at school. "How should human beings relate to the world?" Nejma wondered. "What do we owe to other people when it comes to the environment?"

It was at school where Nejma took the step from looking and thinking to acting. She joined Fridays for Future

France, a regional branch of the global climate strike movement started by Greta Thunberg in August 2018. Sixteen-year-old Greta sat outside the Swedish parliament every Friday, a school day, to demand urgent action on climate change. It angered Nejma that a child was having to sacrifice their education to get society to see the climate crisis for what it is: a crisis.

Nejma began co-ordinating strikes in France, sharing ideas with children in other regions – all the way from the South Pole to the North Pole – and joining together on international protests. For Nejma, she strikes because she cares for the planet and for other human beings.

Nejma hopes that through her actions she has inspired other people to speak up and join the fight. Sharing an analogy between her activism and the workings of nature, Nejma describes how she has planted a seed, and although she may not be able to cultivate the crop grown from the seed, she has made a contribution, ensuring that the Earth is easier to walk on for subsequent generations.

Everybody has the power to create change and Nejma will carry on fighting until her future, and that of many others, is safe. She tells the climate activists who will follow in her footsteps to: "Never stop and never accept things as they are. . . Keep fighting for things and keep getting outraged because that's how you're going to stay powerful."

"I hope for a more EQUAL WORLD – one without racism, sexism, elitism – a world that's kinder to the ENVIRONMENT."

YUJIN KIM

BORN: 2002
ETHNICITY: KOREAN

REDUCE, REUSE AND RECYCLE – Yujin has followed the three 'Rs' since the age of eight. Now in secondary school and an organizer for Youth 4 Climate Action Korea, Yujin is suing the government. Along with eighteen other young activists, they claim South Korea's greenhouse gas emission reduction target is inadequate. If successful, the government would be bound by law to strengthen the country's greenhouse gas emission reduction goals. . . an impressive feat!

Yujin grew up in South Korea and has always had an ecological awareness. After taking small actions to help the planet – from using less water, walking instead of taking the car and eating less meat – Yujin learned that personal change can only go so far, and that government action can make a bigger difference.

When Yujin was briefly living in Australia in 2018, she was invited to a School Strike for Climate event. It was the largest rally that she had ever been to, with 30,000 people taking to the streets to demand action on climate change. When Yujin returned home a year later, she was excited to find a school strike in Seoul, the capital of South Korea, with Youth 4 Climate Action Korea. Yujin made a sign, wrote a speech and showed up to demand change. She then took it a step further and became an organizer.

Youth 4 Climate Action Korea comprises of around 100 teenagers who protest to alert the government to the issue of climate change. Aware that South Korea is the world's seventh-highest emitter of carbon dioxide, in 2019 they demanded stronger climate policies and greenhouse gas emission reduction goals. "We did what we could to make changes by meeting government officials," Yujin said, "But it wasn't enough for political leaders to implement better policies." Time was running out. . .

So, in March 2020, Yujin and eighteen other young activists sued the government, saying its climate policies were violating children's human rights. The lawsuit passed the preliminary review just ten days after they submitted the case.

For Yujin, government decisions define what kind of world today's youth will inherit, so, she says, "It is only fair that our voices are heard and our demands are taken into account in the making of those decisions." What she wants is simple: a healthy planet. It is our fundamental human right, after all.

> *"It is the GOVERNMENT who needs to take responsibility for protecting its people from CLIMATE CHANGE."*

ANDER CONGIL ROSS

BORN: 1997
ETHNICITY: SPANISH, BASQUE, AFRICAN TRINBAGONIAN

ANDER HAS A CONNECTION TO NATURE, and has been interested in birds and photography since he was a child, just like me. He has always felt a desire to leave a place better than how he finds it, so wasted no time in responding to Greta Thunberg's call to action. In 2019, Ander co-founded Fridays for Future Girona, a regional branch of Spain's climate strike movement, and has been demanding global climate action ever since.

A nder was born in Spain and was raised with the philosophy that we all have to do as much as we can to protect the planet. He recalls spending time in the countryside, collecting litter and developing a love for birds. He found birds intriguing as the species are so diverse.

So, it seemed natural to him to study Biology and Environmental Sciences at the University of Girona. It was there, on a cold Friday in January 2019, that Ander and three university friends skipped class to stand in the Generalitat of Girona to help support a movement that they had been following on the internet: Fridays for Future. The week before they had set up Fridays for Future Girona, a regional branch of Spain's climate strike movement, and on that Friday joined students across the world for the first time. They stood together to demand climate action from their political leaders. "Fridays have never been the same since," Ander says.

The first few weeks of strikes were calm, but as news spread on social media, the regional branch began to expand. They gathered more than 100 people, and from there it took off! During a climate strike in March 2019, Ander took to the streets alongside 50,000 people to denounce political inaction on the climate crisis. Ander felt emotional knowing they were all there to fight for a better future.

For Ander, it's important to talk about the consequences that the climate crisis will have. He is worried that Spain is the country in the European Union most vulnerable to climate change, and as such should be one of the pioneering nations in environmental policies. Instead he feels that time is running out and, at this point, the government is not showing any signs of adopting the change that it should.

Ander keeps in his head a phrase that, if we all followed, would combat the climate crisis: "Leave the place better than you find it." So, alongside striking, he continues to take part in clean-up projects in meadows and rivers and on beaches and mountains.

"We have to ACT to SAVE the PLANET before it is too late. It is your lives that are being played with."

ARCHANA SORENG

BORN: 1996
ETHNICITY: INDIAN KHARIA

ARCHANA BELONGS TO THE KHARIA TRIBE, an Indigenous forest-dwelling community living in India. She advocates for tribal communities to have a pivotal role in the climate activism debate, and fights to protect their rights in the face of environmental and land destruction. I share Archana's passion to provide a platform for marginalized voices, and relate to how her climate activism has been shaped by her community's struggles and her parents' effort.

When Archana was growing up in Bihabandh village in Rajgangpur, her grandfather was a pioneer protector of their forest community. He led the forest protection committee, and her father cured illnesses using ingredients from the forest, ways that were learned and passed down the generations. When Archana's father passed away in 2017, she became interested in activism and research as she wanted to preserve her community's knowledge.

Since childhood, Archana's parents told her that to change society she needed to enter politics and influence decision-making processes. So, Archana went to university to study Political Sciences, and started to visit other tribal communities. She saw how similar they were to hers – how they had an environmentally friendly way of living and how nature was deeply embedded in their lives. Archana shares that, for example, her clan name 'Soreng' means 'rock' in the Kharia language – and other clan names include 'Kiro' meaning 'tiger' and 'Dung Dung' meaning 'fish'.

Archana learned that sustainable solutions like cleaning water and organic farming were being carried out, but had not been shared outside of different tribal groups. She started to write about communities' successes and shared their traditions, knowledge and practices.

Archana joined the United Nations' Secretary-General's Youth Advisory Group on Climate Change and started advocating for climate action. "Securing land rights is key to ensuring Indigenous communities can contribute to climate action," she explains.

Archana believes that language must change to demonstrate Indigenous people are conserving nature sustainably. "Indigenous communities need to be the leaders, not victims, of conservation," Archana says. Forest, nature and land embody their identity – nature is their mother, who is taking care of them, and, in return, they must take care of her.

> "GUARDIANS OF THE FOREST are on the verge of EXTINCTION as their RIGHTS are not recognized."

JAVIER CANG

BORN: 1990
ETHNICITY: FILIPINO

OUTDOOR PHOTOGRAPHER JAVIER is a climate activist and storyteller. He shares images online of different wildlife and habitats that have been impacted by climate change. Javier is a keen adventurer and has travelled the world but believes that, when it comes to global environmental issues, the Philippines' voice needs to be louder. That's why he's representing his country as the ambassador of Prince William's prestigious Earthshot Prize.

Javier grew up in the Philippines and his love of nature and spending time outdoors came from his parents. As a young child, he recalls family hiking trips where he started to ask questions like, "Why are there fewer trees in this forest?" and "Why does the mountain no longer have snow?" He learned habitats were changing for the worse and decided to share their stories with the rest of the world. He picked up his camera and hasn't looked back.

After starting a social media account, Javier began to build a platform to show his photographs. His explorations have led him to the top of some of the highest peaks in the Philippines, along with mountains in Borneo, Japan and Mongolia, to name just a few. It is in these places where Javier uses his camera lens to show the face of climate change – broadcasting to over 40,000 followers.

Using his platform for positive change, Javier is an ambassador of the Earthshot Prize. Launched by Prince William, the Duke of Cambridge, each category is dedicated to a different environmental issue, such as building a waste-free world, protecting and restoring nature, cleaning our air and reviving our oceans.

As the only Filipino ambassador, Javier felt it was important to bring the environmental concerns in his country to the forefront. Due to industrial waste and cars, Manila, the capital of the Philippines, suffers from air pollution from fossil fuels, which affects 98 per cent of the population. But clean, healthy air is within reach by making renewable energy accessible – something Javier shared when narrating a film for the Earthshot Prize.

Javier hopes to inspire the next generation to drive change. "The most powerful tool that anyone has is their voice," he explains. For Javier, we're incredibly privileged to have the chance to enjoy the planet, marvel at the mountains and swim in the oceans. . . but it is a privilege. And it is a privilege that needs protecting.

"By sharing PHOTOGRAPHS and telling STORIES, I hope to build a society that is more in tune with NATURE."

SCARLETT WESTBROOK

BORN: 2004
ETHNICITY: BRITISH—KASHMIRI

SCARLETT VOLUNTEERS AT TEACH FOR FUTURE, a student-led project which campaigns to change the education system so that it centres around the climate emergency. Aware that students are growing up in a world deeply affected by climate change, both Scarlett and I believe they deserve to be taught to understand the impacts and how they can help.

A t the age of thirteen, Scarlett became the youngest person in the world to gain an A-level in Government and Politics. The qualification focused on education and climate policy and provided her with a solid foundation for the campaigning work she does today.

Scarlett attended school in Birmingham, but wasn't taught about one of the most important issues facing her generation: climate change. Shockingly, only 4 per cent of students feel that they know a lot about climate change – a statistic Scarlett is passionate about improving. She advocates for climate change education to be established in every subject as, Scarlett says, "The effects of climate change will affect everyone."

However, Scarlett is aware that some people and places will be more affected, for example those from poorer countries. This is a result of the restriction on a nation's economic growth during colonization (when one country takes control of another). This, Scarlett believes, shows the importance of decolonizing the education system. What Scarlett means by this is that students should be able to question the viewpoint of the information being taught to them.

As part of Teach the Future, Scarlett works with other students to advocate for an education system that represents all, not just the most privileged. Through her campaigning, Scarlett breaks down traditional access barriers so that everyone is involved in the conversation. Representation is vital in achieving this, as Scarlett says, "We may be able to discriminate against each other, but the impending threat of climate change will not. It is imperative that we put aside our prejudices to unite."

Which is why Scarlett's work is so important: by refocusing the education system, students can develop the skills to think more critically about their future and develop the agency to positively influence it.

> "We are living in an age where YOUNG PEOPLE should be more INFORMED about CLIMATE CHANGE than ever before."

DAVID ESTEBAN

BORN: 1999

ETHNICITY: COLOMBIAN, AMERICAN MUISCA

GROWING UP IN THE ANDES MOUNTAINS IN COLOMBIA, David lived a healthy, sustainable life with his family. But when a soft drinks manufacturing company took the community's water, David's life changed forever. He became a climate activist, and fights for a world where big corporations can no longer harm the planet or the people who inhabit it.

D avid first became aware of the dominance of big corporations when he was seven years old. A soft drinks manufacturing company came to the mountains near David's home in Colombia and built a water extraction plant which drained the river – taking up to a million litres of water every day. Latin America has 31 per cent of the world's fresh water, the largest of any region. Since fresh water appeared to be plentiful, David's government wasn't focused on its conservation, and his community wasn't educated about how to protect their supply.

So, David and his family had to relocate to find fresh water. They needed it for drinking, as well as for their crops and animals.

With their natural water sources drained, people had to walk a long way to gather water, with some having no choice but to buy plastic bottled water. The big corporation grew larger as it persuaded people in other countries to buy water in plastic bottles too. People didn't know that this water had been taken from communities around the world, so David began to speak out.

He shared his story on Greenpeace panels and attended marches to raise awareness of the plight. "By taking away our water supply, they took away our way of life," David says. It was shocking to people that governments weren't providing clean water, but a large corporation had taken control of one of the best water sources.

Alongside his environmental work, David spoke out about the impact on his mental health. David and his family were forced to move to the city as a result of the water extraction plant, leaving his grandparents behind. He struggled to adjust and experienced depression from, he shares, "leaving the freedom and life I knew behind".

Now residing back in the mountains, David's work focuses on his farm. He organizes camps and teaches young people in his community how to produce their own food and live off the land. This sustainable way of life puts less strain on the environment, and David dreams of protecting the planet from its current dependence on big corporations for food and water.

"Humans were made to PROTECT Mother Earth and live in HARMONY."

SAGAR ARYAL

BORN: 1995
ETHNICITY: NEPALESE

SAGAR STARTED CLIMATE CHANGE CAMPAIGNING when he was ten years old. He read an article about the melting ice on Mount Everest, a mountain in his home country of Nepal, and decided to act. Sagar's tree planting project is an important, physical act. I plant fruit trees in my village in Bangladesh and so I share Sagar's dream to combat climate change through reintroducing nature.

As a child, Sagar understood how badly Nepal was changing as a result of climate change and was upset that those around him didn't understand the issues. Stemming from a belief in 'Education for All', he started a reading group in Kathmandu, the capital of Nepal. With co-operation from his family, he collected over 3,000 books in a couple of weeks. The club was both a study centre and a discussion forum, and every Saturday the reading group debated a range of issues, including global warming. By teaching others about the climate crisis, Sagar knew they could work together to make a difference.

Knowing that when trees grow they absorb and store the carbon dioxide emissions that drive global warming, Sagar moved to Germany to work on a project for Plant for the Planet. The scheme aims to plant a trillion trees around the world over the next ten years. The hope is to capture an entire third of human-made carbon dioxide and bring transparency to individual planting efforts.

The Global North, in countries like Germany, funds the planting of trees in the Global South, in countries such as Nepal. By doing so, the project creates millions of jobs and also increases wealth in the Global South, as all of the money raised goes directly to the tree planters.

Sagar's work helps the environment, and also creates opportunities and platforms so everyone can take action. Sagar sees educating the next generations as integral to his activism, and wants to engage all voices in the climate change discussion – especially voices from the Global South.

Sagar's aware that climate change is already difficult because it's a political topic, which makes it even harder for Indigenous people and people of colour to engage and be heard, and yet their perspective is critical. "Every culture is different, and every society has different ways of integrating with the planet," he acknowledges. Sagar believes everyone can take responsibility for society and take action to make it a better place.

"I will leave the planet a BETTER, HEALTHIER place so that children growing up don't have to live through the same worry."

44

TAYLOR CLARKE

BORN: 1998
ETHNICITY: AUSTRALIAN FIRST NATION BURRAGORANG VALLEY GUNDUNGURRA

TAYLOR COMES FROM THE BURRAGORANG VALLEY IN THE BLUE MOUNTAINS in New South Wales, Australia, where she works as an Aboriginal Culture and Heritage Educator. Taylor shares her experiences with locals to foster a more connected relationship with Indigenous groups. Championing the 'Give a Dam' campaign to stop the raising of the Warragamba Dam wall, Taylor fights to prevent the destruction of wilderness rivers and Indigenous culture.

Fifty years before Taylor was born, the first walls of the Warragamba Dam went up, flooding the Burragorang Valley and forcing the Gundungurra people from their homes. Taylor's great-great-grandmother refused to leave, and her sons had to carry her out. Warragamba Dam currently provides water to 4.5 million people in Australia, but at a significant cost to Taylor's community.

In the 1980s, the government raised the wall again, and they are planning to increase its height once more. . . which would change the Blue Mountains forever. What could be lost by raising the wall? "Critically endangered plants and animals, hundreds of art and burial sites, as well as one of the largest intact Aboriginal songlines in this part of Australia," Taylor explains.

The new dam proposal started three years ago, just after Taylor started university. The consultation with Aboriginal people was non-existent, so Taylor reached out and partnered with a foundation that was willing to listen. Taylor provided her unique perspective and helped to shape the campaign. But even Taylor admits, "Part of the problem is that we don't know the full breadth of what will be lost. When the dam was made, people moved out of the area and they took with them their knowledge."

The passion for her family history feeds into her work as an Aboriginal Culture and Heritage Educator. Taylor speaks in schools to make sure Indigenous culture is not forgotten. She is also a member of the Nature Conservation Council in New South Wales, providing an environmental voice for her community.

Taylor dreams of having a democracy in Australia that has a place for Aboriginal people and all unrepresented communities. But, as Taylor knows, "Nothing worthwhile is easy," calling to others to, "Push against the structures that are built to destroy the things you care about to make real change."

> "Aboriginal people are real people. . . our history and culture must be equally VALUED to protect our NATURAL HERITAGE."

ISAO SAKAI

DREAM: RESHAPE OUR FUTURE SUSTAINABLY

BORN: 2001
ETHNICITY: HONSHU JAPANESE

ISAO IS A FOUNDING MEMBER OF FRIDAYS FOR FUTURE TOKYO, urging the Japanese government to act to fight the climate crisis. He organizes demonstrations against corporations that act against the environment's best interests. Impressively, Isao has also contributed to the 'Let's Divest' campaign at 350.org Japan as a core volunteer, highlighting how Japanese banks invest in coal power plants worldwide and use consumers' money to accelerate the climate crisis.

Isao grew up in Tokyo, the capital of Japan, thinking that climate change affected other countries, but not his own. But after taking an environmental science course at school during a year abroad in the United States, he realized that his own future was at risk. Extreme weather affects Japan and, Isao honestly shares, "With the typhoons, droughts and other climate disasters, I thought I can reasonably lose my own future."

It became clear to Isao how much Japan contributed to the climate crisis as the world's fifth-biggest emitter of greenhouse gases. It's Isao's belief, therefore, that Japan has a responsibility to protect its people.

So, when Isao returned home, he joined the 'Let's Divest' campaign at 350.org Japan and began campaigning against Japanese banks investing in coal power plants — as digging up, transporting, processing and burning coal is a huge contributor to greenhouse gas emissions. Shortly after, he co-founded Fridays for Future Tokyo, a regional branch of Japan's climate strike movement. Isao drew thousands of activists to the streets of Tokyo, calling on city officials to revise their emission reduction goal.

When Isao first started campaigning, his family and friends were sceptical as protesting is seen as radical in Japan. There can be a stigma attached to people who participate in activism, but Isao continued to fight. Thankfully, his family understood, with his mother starting to use renewable energy. After six months, thirty of his fellow students had also joined Isao at the strikes.

Isao is passionate that strikes can change the future, but feels anger is not a good weapon to use. He thinks that people should not waste time criticizing each other, but be tolerant of the differences in approaches. Isao believes that the future is ripe for reshaping, and at the moment we are wasting the planet's resources at dramatic rates: "We are mad at the systems that are getting worse, but we need to make the change that we want to see."

"We have to RESHAPE the future to protect the planet. Let's be DIFFERENT. Let's do BETTER."

NAILA SEBBAHI

BORN: 1998
ETHNICITY: MOROCCAN

NAILA BEGAN HER ACTIVISM JOURNEY AS A HUMAN RIGHTS CAMPAIGNER, and campaigned for social justice and human rights during the migrant refugee crisis in Europe. After becoming aware of the link to the climate crisis, Naila co-launched a global environmental movement in Belgium to raise awareness and inspire change. Naila believes we must have a national emergency plan which builds a more sustainable, fairer world that works for everyone.

N aila was born in Morocco and moved to Belgium as a young child. In 2015, she started campaigning for social justice and human rights. During this time, over five million refugees and migrants had reached Europe, after undertaking journeys from Syria, Iraq, Afghanistan and other countries to escape conflict and climate change. Their plight struck a chord with Naila as a daughter of immigrants.

While organizing large-scale protests to call for racial justice and a fairer migration policy, Naila learned about the link between migration and the environment. Climate

migrants are people who are forced to leave their homes due to changes in their environment. They are the most adversely affected by climate change, but they are the least responsible and the least equipped to deal with it.

So, in 2018, Naila joined a call to action and helped launch the Belgian branch of Extinction Rebellion, a global environmental movement demanding national emergency plans, developed by citizens themselves, with support from scientists. A plan, she hopes, that would manage a socially just transition from fossil fuels towards renewable energy.

Fossil fuels include crude oil and natural gas, and when they are burned, they release greenhouse gases which contribute to climate change. "Seeing how stubborn we've been for the last four decades of warning, the climate crisis shows up as an inevitable lesson for humanity," Naila explains, "But if we wish to land with the least possible damage in the fairest possible way, healing ourselves from our addictions, including fossil fuels, would be a very good start."

Alongside her work with Extinction Rebellion, Naila explores new ways of practicing democracy and building a regenerative culture of being and co-existing, not "in", she stresses, but "as an integral part of" the natural world.

"We're demanding a NATIONAL EMERGENCY PLAN which stops the extraction of FOSSIL FUELS."

CARLOS MANUEL

BORN: 2002
ETHNICITY: FILIPINO

AFTER WITNESSING A TROPICAL CYCLONE AT THIRTEEN YEARS OLD, Carlos became concerned about how the climate crisis was damaging the marine environment in Palau, Oceania. He graduated from secondary school and started volunteering at the Coral Reef Research Foundation (CRRF). By analysing the impact of climate change and pollution on the oceans, Carlos acquired knowledge to conserve and safeguard the waters around his home and the animals which inhabit it.

Born in the Philippines, Carlos moved to Koror, one of the islands that is part of Palau, when he was eight years old. A few years later, an island north of Palau was hit by a tropical cyclone, forcing the inhabitants to move. Carlos realized there was a link to climate change; storms receive their energy from the ocean and the warming oceans make tropical cyclones more likely.

Carlos recalls swimming in the ocean every day with his friends, until, "Almost overnight, the water started getting hotter," he explains. The sea level began to rise, resulting in waves crashing into homes and areas where Carlos and his friends played together. The changes in the water impacted the marine wildlife and threatened the beautiful coral reef. Carlos felt a desire to act.

So, he became a volunteer at the Coral Reef Research Foundation (CRRF).

At the CRRF, along with scientists, Carlos helped to monitor the warming waters which is key to tracking global warming. He learned that in Palau's Jellyfish Lake, the rise in water temperature caused a change in the nutrients in the lake, affecting the food chain and, ultimately, the survival of the juvenile jellyfish. In the last few years, ongoing monitoring and conservation work has indicated that the jellyfish populations are finally rebounding. "What else could be achieved?" Carlos wondered. "It's now time to bring the conversation to the rest of the world."

At the age of seventeen, Carlos joined a group of sixteen youth activists in petitioning the United Nations (UN) to combat climate change. Speaking at the UN headquarters, Carlos said, "Small islands are the most vulnerable countries to be affected by climate change."

Today, Carlos wants to encourage young people to keep trying to protect the planet because, he says, "If everyone works together, we are unstoppable." Carlos cares deeply for future generations, and galvanizes them to protect what is most important to them. And for Carlos, that will always be the oceans.

"Our oceans and its marine life are RAPIDLY DETERIORATING due to the CLIMATE CRISIS."

CAITLYN BAIKIE

BORN: 1992

ETHNICITY: INUK, CANADIAN

CAITLYN HAILS FROM A SUBARCTIC REGION OF CANADA CALLED NUNATSIAVUT – which means 'Our Beautiful Land'. Starting her environmental activism at the age of fourteen, Caitlyn advocates for global awareness of the human impacts of climate change, particularly on Indigenous people. Her work focuses on an all too common problem of important climate issues for the Indigenous community being largely overshadowed and ignored in the media.

Raised in the Inuit community of Nain, Nunatsiavut, with a population of just over 1,400, Caitlyn was out on snowmobiles with her family from the age of five months, and recalls how her small community helped foster a desire to advocate for the planet. At fourteen years old, she went on a work trip with her father to Torngat Mountains National Park and learned from researchers who were studying climate change in the area.

Ever since, Caitlyn has had a growing sense of awareness of the human impacts of climate change, particularly on Indigenous people. Inuit are seeing significant changes in their climate patterns, with a lot more snow and less ice. For Inuit, culture is an oral tradition and they pass on their cultural knowledge by being out on the land and connecting with nature. However, this has been negatively impacted in recent years with the migration patterns and populations changing, with more non-native species in their area, including moose and foreign birds.

Already, the Arctic is warming at twice the global average, and, in Caitlyn's view, the media talk about the disappearing Arctic ice, but there is almost no conversation around Arctic people and their changing livelihoods. As Caitlyn points out, "We contribute the least amount of what's causing a lot of climate change, but yet we are the most impacted by that."

Caitlyn has published articles on the Arctic, presented at conferences around the world to raise awareness for their plight, and is part of a research project focused on the impact of climate change on Indigenous people's mental health.

Caitlyn says, "Everybody should be open to different world views. If you're not Indigenous, get to know Indigenous people. If you're Indigenous, engage with research." If we work together with open minds, we are in a strong position to create lasting change.

> "I hope for Indigenous VOICES and EXPERIENCES to no longer be SILENCED and our KNOWLEDGE to be RECOGNIZED."

AMMR MOHAMED ABDEL SAYED

BORN: 1997
ETHNICITY: EGYPTIAN, SUDANESE

AMMR UNDERSTANDS CLIMATE CHANGE IS A GLOBAL EMERGENCY that goes beyond national borders, as he was born and raised in Italy by parents from Egypt and Sudan. At fourteen years old, Ammr became interested in social issues, educating himself at workshops, and then ecological issues while studying at university. He later joined a regional branch of Italy's climate strike movement, Fridays for Future Torino, and organized local youth strikes, taking to the streets in his fight for the planet.

Ammr was inspired to become a climate activist when he learned the planet was in danger as a result of climate change. In 2013, Egypt, where Ammr's father is from, was taken over by a military coup (a sudden overthrow of a government) which intensified pressures on both human rights and the climate crisis. Ammr was only sixteen years old, but he remembers wanting to help.

With the belief that everyone can make a difference, Ammr joined a regional branch of Italy's climate strike movement, Fridays for Future Torino. He missed school to unite with other students and youth activists from Turin to demand action from political leaders to limit climate disaster. Ammr plays a significant role in organizing the regional strikes, but sees himself as just a small part of the wider global climate justice movement.

The Fridays for Future strikes have had a huge impact across the world, yet the Covid-19 pandemic in 2020 forced the movement to adapt. Taking the fight online, the strikes continued virtually, however once regulations eased, Ammr was back protesting on the streets of Turin. "The coronavirus pandemic made clear the contradictions of our economic and social system," Ammr says. "It forced us to treat any emergency situation by listening to the science. Despite this, the climate crisis continues to be ignored by the government."

Although we may think about specific global issues, like health and the environment, as separate challenges, the reality is far from it. In Ammr's view, they are deeply connected, and our understanding and response should be shaped accordingly.

From what Ammr has learned studying Social Economics at university, he believes in the importance of collaboration, asserting, "We must connect political, environmental, social and economic issues, and work together as one."

"In SEVEN YEARS, the planet will reach a point of NO RETURN. We must ACT NOW."

GAURI SHUKLA

BORN: 2002
ETHNICITY: INDIAN

ON A VISIT TO INDONESIA, GAURI SAW A FIRE ENGULF LAND TO MAKE WAY FOR PALM OIL PLANTATIONS which created huge amounts of haze. On her return home to Singapore, she founded Students of Singapore Against Haze. Gauri works to educate people about the causes and health hazards of haze, alongside petitioning companies to switch to using sustainable palm oil – which is important to protect the forests and the wildlife that live there.

In 2015, Southeast Asia was hit by a terrible haze crisis and in Singapore, where Gauri lives, children had to stay inside to protect themselves from the unclean air. Gauri was only thirteen years old at the time, but she wanted to understand where the haze originated. She discovered that haze particles were coming from forests being burned in nearby Indonesia and Malaysia as farmers cleared vegetation for palm oil plantations.

Indonesia and Malaysia make up over 85 per cent of the global supply of palm oil. Palm oil is a vegetable oil that comes from the fruit of oil palm trees, and is in nearly everything from pizza and chocolate to shampoo and toothpaste. It surprised Gauri to learn that her consumption of palm oil was linked to the haze, and she wondered if other people knew. . .

A year later, Gauri saw first-hand the destruction on a trip to Bukit Lawang, Indonesia. "I hated what I saw. Lush green forests had given way to oil palm plantations and stacks of logs. The whole place was shrouded in thick smoke," Gauri recounts.

So, Gauri set up an organization, Students of Singapore Against Haze, to raise awareness about the risks of haze and put pressure on companies who contribute to creating it. It's not as easy as stopping people using palm oil, as millions earn their income from it. Instead, Gauri reached out to companies to ask them to buy palm oil from farmers which adhered to the global standard set by the Roundtable on Sustainable Palm Oil.

Benefitting from the reach of social media, Gauri petitioned the makers of Singapore's iconic curry puff snack to switch to sustainable palm oil. And the petition drew more than 8,000 signatures! Gauri's goal is for Singapore to pioneer a sustainable food industry, and she is working hard to create an open dialogue. A solution can be found only when, Gauri says, "All of us work together".

> *"Irresponsible production of PALM OIL is causing widespread rainforest DESTRUCTION."*

ERISVAN BONE DE SOUSA SILVA

DREAM: INDIGENOUS VISIBILITY AND INCLUSION

BORN: 1996
ETHNICITY: GUAJAJARA BRAZILIAN AMAZONIAN INDIAN MARANHÃO

ERISVAN IS A JOURNALIST AND CLIMATE ACTIVIST amplifying the voices of Indigenous people across Brazil. When he was twenty-one years old, he founded Midia India, a communication network which provides a platform for Brazilian Indigenous people to rise up against the lack of visibility in the media. I stand with Erisvan in his campaign to show that Indigenous people should be the protagonists of their history.

Hailing from the village of Lagoa Comprida, on Araribóia Indigenous land in Brazil, Erisvan has been fighting against land encroachment for most of his life. Land encroachment is when a person's territory or rights are intruded upon, and Erisvan campaigns for Indigenous land to be marked on maps to make the practice more difficult. For this to happen, Erisvan uses his platform as a journalist to share Indigenous peoples' stories to show how they are guardians of the forest.

Erisvan recalls becoming involved in environmental activism after being inspired by his Indigenous sister and collaborator, Sônia Guajajara, an activist also defending the rights of Indigenous people. Erisvan acknowledges that it is necessary to listen to the wisest leaders and chiefs. "We need to follow the right path and respect their cultural and diverse wealth," Erisvan explains.

When Erisvan set up his organization Midia India, his aim was to create a communication network that strengthens the voices of Indigenous people through an independent media. The project respects each community and is based on a collaborative approach, connecting and empowering young Indigenous people from Brazil. Communicators cover issues related to the environment, such as forest fires in the Amazon, invasions by miners and loggers, as well as fighting land encroachment.

Reaching 305 ethnic groups that speak 274 languages, Midia India's valuable contribution won the Joan Alsina Prize for Human Rights in 2020 for, "Disseminating the reality of the Indigenous people of Brazil at a time of many threats," including the Covid-19 pandemic and the destruction of forests as a result of climate change.

Erisvan has real hope for the future and often thinks about the next generations. He acknowledges they must fight for democracies which respect everyone, feeling strongly that "Hate must end and love must always prevail."

> "Indigenous people should have a VOICE and VISIBILITY in their struggle."

INDEX

If you've been inspired by these incredible campaigners, here are some ways to cultivate your own dream:

EDUCATE YOURSELF
Discover which climate-related issues you care about. Reach out to those who are engaged and listen to what they have to say.
un.org/en/climatechange

ENGAGE WITH SOCIAL MEDIA
Take action digitally. Social media is a powerful resource for education, knowledge sharing and making your voice heard.
greenpeace.org/international/act

MAKE YOUR GOVERNMENT WORK FOR YOU
Lobby your government to support your campaign by writing letters or sending messages via social media. When you are eligible, make sure you vote.
parliament.uk/get-involved

HELP YOUR COMMUNITY
Tackle climate change in your own community, from recycling to volunteering. Local action is just as important as global action.
tnlcommunityfund.org.uk/insights/community-action-for-the-environment

JOIN A CLIMATE STRIKE
Go on strike! By attending a climate strike, you will increase its effectiveness; change happens when we stand together.
fridaysforfuture.org